Rain
A Great Day for Ducks

by Jane Belk Moncure
illustrated by Joy Friedman

Created by

Distributed by CHILDRENS PRESS®
Chicago, Illinois

Grateful appreciation is expressed to Elizabeth Hammerman, Ed. D., Science Education Specialist, for her services as consultant.

CHILDRENS PRESS HARDCOVER EDITION
ISBN 0-516-08120-9

CHILDRENS PRESS PAPERBACK EDITION
ISBN 0-516-48120-7

Library of Congress Cataloging in Publication Data

Moncure, Jane Belk.
   Rain : a great day for ducks / by Jane Belk Moncure ; illustrated by Joy Friedman.
   p. cm. — (Discovery world)
   Summary: Illustrates why ducks and other living things like rain. Also describes the water cycle and how to make your own raindrops.
   ISBN 0-89565-553-5
   1. Rain and rainfall—Juvenile literature. [1. Rain and rainfall.] `I. Friedman, Joy, ill. II. Title. III. Series.
QC924.7.M66   1990
551.57'7—dc20   89-24010
                                                           CIP
                                                           AC

©1990 The Child's World, Inc.
Elgin, IL
All rights reserved. Printed in U.S.A.

1 2 3 4 5 6 7 8 9 10 11 12 R 99 98 97 96 95 94 93 92 91 90

Rain
A Great Day for Ducks

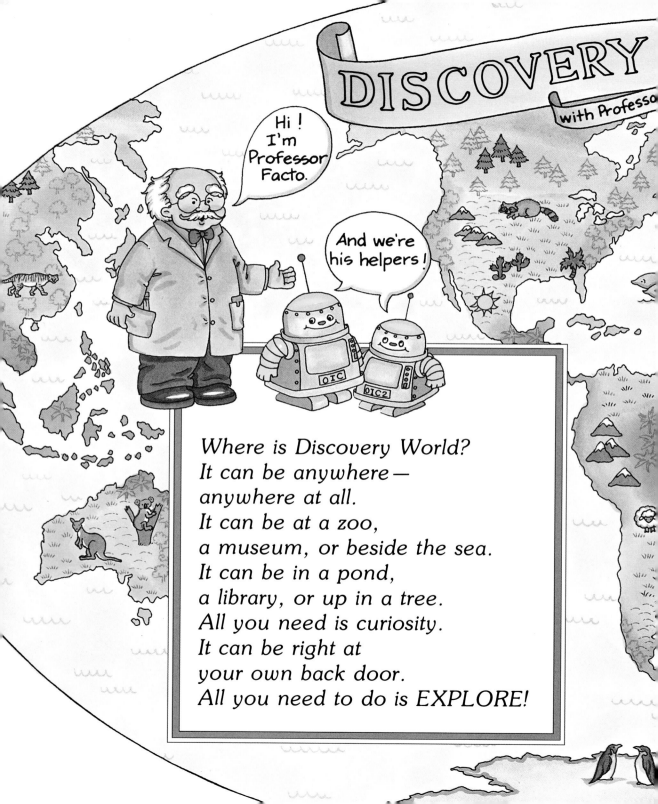

Splish, splash! It's raining all around.
Who likes rain? Ducks do, that's who.

Do they get wet? Not at all. Ducks have rainproof feathers with oil on them to keep water away from their bodies.

Who else likes rain? Frogs do, that's who. Their thin skin needs to stay wet, or it will dry out and they will die.

Trees like rain too. Rain helps trees grow tall.

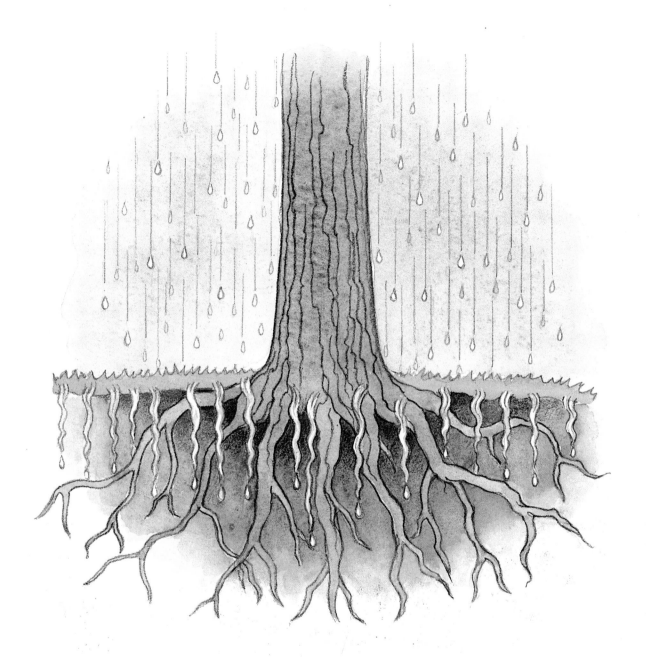

If you could peek under the ground, you would see raindrops dripping down to the roots. Trees get water from their roots.

Rain brings water to all living things.
It helps grass grow for cows and sheep.
It helps flowers bloom for butterflies and bees.

Rain helps the farmer grow corn and wheat . . .

and other good foods we eat each day.

If it never rained again, the ponds and lakes would slowly dry up.

The ducks would have nowhere to swim, and the frogs would have nowhere to live.

But as long as rain falls from the clouds, ponds will stay full, and the frogs and ducks will stay happy.

Sometimes rain sprinkles down from the clouds in a light shower. *Plip, plop!*

Sometimes rain pours down in a noisy thunderstorm. *Crash! Bang!*

Sometimes rain falls in a steady downpour. And after it stops, everything smells clean and fresh.

Birds hop about, looking for earthworms.

Water sparkles on the leaves and sometimes looks like tiny jewels on a spider's web.

And if you look up at the sky, you might see a big, beautiful rainbow!

Rain makes the world a beautiful place.
And it gives plants, animals, and people
the water they need to live.

That's why a rainy day is a great day for ducks and a great day for you!

NOW EXPLORE SOME MORE WITH PROFESSOR FACTO!

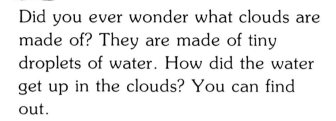

Did you ever wonder what clouds are made of? They are made of tiny droplets of water. How did the water get up in the clouds? You can find out.

1. After the next rainfall, find a puddle on a sidewalk or driveway. Draw a circle around the edge of the puddle with a piece of chalk.
2. Come back to check on the puddle the next day. What happened?

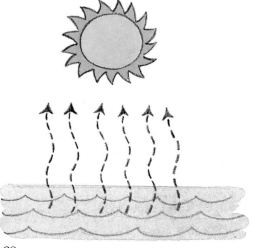

When the sun heats up water, the water turns into a gas called **water vapor** and rises up into the air. We can't see water vapor, but it is in the air all around us. When water vapor cools, it changes into water droplets and forms clouds!

You can make your own raindrops! Here's how:

1. Fill a clear, plastic glass with ice.
2. Put the glass on a sunny windowsill.
3. Come back in 15 minutes to check on the glass. What do you notice on the glass? Where did the water come from?

As the warm air outside the glass meets the cold glass, water vapor in the air turns into tiny drops of water which form on the glass. They look just like raindrops as they slide down.

Did you know that the earth uses the same water again and again? Here's how:

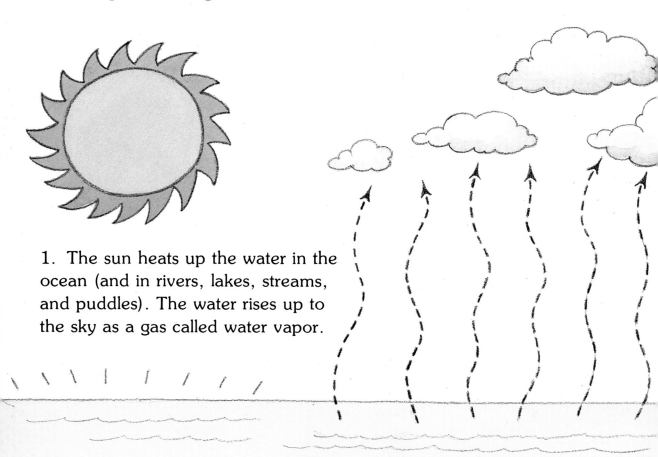

1. The sun heats up the water in the ocean (and in rivers, lakes, streams, and puddles). The water rises up to the sky as a gas called water vapor.

2. The water vapor cools and changes to water droplets. The water droplets form clouds.

3. The water droplets in the clouds get bigger and heavier until they finally spill out as rain.

4. Rainwater flows back to the ocean in rivers and streams. The cycle goes on and on!

INDEX

clouds, 18, 20, 28, 31
condensation, 29, 31
ducks, 6-7, 17
evaporation, 28
food, 14-15
frogs, 9, 17
rainbow, 25
showers, 20
thunderstorms, 21
trees, 10-11
water cycle, 30-31
water vapor, 28, 29, 30, 31